The Native Lowland Sheep of Galway and Roscommon: A History

– P. L. Curran –

Peace is essential to successful stock-breeding
[page 16 in Wm. Fream's 1900 edition of W.Youatt's The Complete Grazier etc., Crosby Lockwood]

MMI

Printed in the Republic of Ireland by
A & J Print Dunshaughlin Ltd., Greenpark, Dunshaughlin, Co. Meath
Tel. 01-8256402, Fax: 01-825 6009
E-Mail: ajprint@eircom.net

"Cúilín"
Macetown
Tara
Co. Meath
Republic of Ireland

A.D. 2001

The Native Lowland Sheep of Galway and Roscommon: A History

ISBN: 0-9541847-0-X (Hardback)
ISBN: 0-9541847-1-8 (Paperback)

The author wishes to acknowledge the sources upon which this account is based. All sources are acknowledged gratefully, including those on which copyright has lapsed. Without them a substantial account would not have been possible.

The author appreciates the help and advice given by Mr. Tom Sice and Mr. Tom Murphy, Joint Honorary Secretaries of the Galway Sheep Breeders' Society, and now (respectively) Secretary and Assistant Secretary of the Galway Sheep Breeders' Association Ltd.

Published by Patrick Leonard Curran

Tom Sice (right) presents the Galway Sheep Breeders' Cup to John Joe Steede for his pen of ewe hoggets, 1997

Abbreviations Used in the Text

ACOT	An Chomhairle Oiliúna Talmhaíochta
AFT	An Foras Talúntais
AGM	Annual General Meeting
CMA	Cyclopedia of Modern Agriculture, 1910
DAFF	Department of Agriculture, Food and Forestry.
DNA	Deoxyribonucleic acid
E.U. or EU	European Union
FAO	Food and Agriculture Organisation of the U.N.
GCCA	Galway County Committee of Agriculture.
GSBS	Galway Sheep Breeders' Society
IGRCT	Irish Genetic Resources Conservation Trust Ltd.
IPGRI	An international research centre that coordinates and administers a Europe wide network of genetic resources
IRBS	Irish Rare Breeds' Society
RDS	Royal Dublin Society
REPS	Rural Environmental Protection Scheme.
RFB	Roscommon Flock Book
STAR	A Committee of the EU that administers the REPS and other programmes

Acknowledgements

The following are the sources of photographs used in the book:
Teagasc: Photos on pages 9, 16 and 71.
Conroy's Photographic Studio, Cross Street, Athenry: Photo on Title Page and those on pages 3, 36(2), 46(2), 47, 48, 63, 64 and 70(1)
Alan Cheese: Photo on page 70(2)
Leicester Longwool Sheep Breeders' Association: Photo on page 19(2)
Fream's 14th edition of Youatt's The Complete Grazier 1900,
Photo on page 19(1) and 20(1)
The Cyclopoedia of Modern Agriculture 1910: Photos on page 28
The Roscommon Flock Book 1896 - 1906 (these were the early editions): Photos on pages 20(2), 29, 30 and 36(1)

List of Illustrations

2000. 3: Martin Fallon, owner of this champion hogget ram, being congratulated by Seamus Cuddy, Ulster Bank (sponsor), at the 1997 GSBS Show and Sale. 4: Best Wool Class, 1997 GSBS Show and Sale, L. to R.: John Maher (Monivea) holding JMM5, Seamus Cuddy (Ulster Bank), and Michael Conneely (Monivea) holding MCM9.

Page 48: Winners in the ram lamb class at the GSBS Show and Sale, 1997: L to R: Michael Conneely (Monivea), Joe Healey (Athenry), Seamus Cuddy, Ulster Bank (sponsor) and Tom Murphy (Ballinasloe) with the latter's champion, MMN1, bought by Martin Davey, Charlestown, Co. Mayo.

Page 63: 1: Owners (with winning rams) at GSBS Show and Sale 1996. L to R: Tom Sice (with PSL6), Tom Murphy (with MML8), Martin Fallon (with BFL4), Paddy Sice (with PSL3), and Bernard Fallon (with champion ram, BFL6, bought by Joe Casey)

2: Henry Hughes, wool expert, examines Joe Casey's prize-winning ram for wool quality in RDS Class at Gurteen Agricultural College, 1997. L to R: Gerry Nolan (Ardrahan), Michael Conneely (Monivea), Joe Casey (Ballyvaughan), Henry Hughes.

Page 64: 1: Children, Michael and Jacinta Higgins (Athenry) with L to R: Paddy and Tom Sice (Athenry), Michael Maher (Monivea), Seamus Cuddy (Ulster Bank, sponsor), Joe and Jack Casey (Ballyvaughan) and Martin Fallon (Tuam), 1997. 2: Young Galway ewe attends to her new born lamb at Tara, 2000.

Page 70: 1: Joe Casey (left), Chairman of the Galway Sheep Breeders' Society, receives trophy for his prize-winning ram from the staff of Channelle Veterinary, with judge, Pat Loughnane (right).2: This Galway ram was Longwool breed champion at the RBST Show at Stoneleigh (Warwickshire), 2000

Page 71: Traditional Sheep Cock, Kilmaine (P.J. Daly[i], 1966)

Traditional póirín restricting cattle but not sheep (idem)

PREFACE

The Heritage Council failed to provide support for this publication under its 2001 Publications Grant Scheme. It is an objective of that body "to promote pride, awareness and understanding of the national heritage". The completed manuscript was submitted to the Heritage Council on 14th May 2001 and an assessor's report was sent to the author on 11th July. All 273 words of the assessment are set out below in the form in which they were received. The author of this book has inserted brief comments in italic lettering after each of the four paragraphs that make-up the "assessment". Readers are invited to judge the validity, or otherwise, of an assessment which had the effect of blocking the award of a grant of 50% of the costs of printing and binding.

The "Assessment" *[with comments]:*
The Native Lowland Sheep of Galway and Roscommon: A History
The stated aim of the book "is to draw attention to the need for this country to play its part in the global effort to stem erosion of Earth's genetic diversity of livestock". There are references to genetic conservation in the introduction (where the standard reasons for conservation are listed) and in the final chapter, but in practice the book does not deal with the subject matter.
[This latter assertion simply is untrue. The book deals with the history of the Roscommon Sheep Breeders' Association and of the Galway Sheep Breeders' Society and with related matters. It points to the extinction of the Roscommon breed and to the recent decline in genetic diversity within the diminished population of pedigree Galway sheep.]

It appears that the book relies heavily on quotations from other sources, and makes only limited attempts to fill the gaps between the quoted material. In this context, the subject matter does not cover an unexplored topic, nor does it show original research. The text lists breed standards, details of flocks, and show/sale results from information in other publications. However, it is the first time references to both breeds have been drawn together in one volume.
[Yes, quotations from acknowledged sources are drawn together in the book and linked by appropriate comments to provide a story in

chronological sequence. There are no "gaps". No other book has explored the topic. The author requested a definition of "original research" from the Heritage Council on 15th July 2001; no such definition was provided. Parts of the book are quite original. For example, many of the photographs and the part that draws attention to the State-funded research with Galway sheep in 2001, which (if continued, unaltered) will seriously reduce genetic diversity within the already diminished pedigree population.]

Enhancement of awareness and understanding of Ireland's heritage could be achieved by a deeper study of the native sheep from which both the Roscommon and Galway evolved. This is the Irish part of the story (compared to the later developments which rested on importations of British sheep to which a significant chapter is devoted). Native sheep are the important element of genetic diversity.
[The "native sheep" to which this paragraph refers have been extinct for more than a century. The author has not found any significant literature on such sheep. This paragraph is without substance or meaning as an "assessment"]

Some sections are dealt with in a full and interesting way. In particular, the chapter on The Roscommon Sheep Breeders' Association, and the section on the decline of the Roscommon in competition with the Galway, were well described. However, the section on the decline of the Galway from 350,000 ewes in 1962 to 300 ewes in 1994 did not explain the causes of the drop in population.
[The last sentence refers to two related but distinct sheep populations. To suggest that the large commercial population of non-pedigree Galway sheep declined after 1962 to become the small pedigree population in 1994 is simply incorrect. To suggest that the author should explain an assessor's misinterpretation of facts that are set out clearly in the book has no rightful place in an official anonymous "assessment"]

[Conclusion: Three of the paragraphs of the "assessment" have little meaning. The contents of the second paragraph form 28% of the "assessment" and give scope for some debate. There was no debate.]

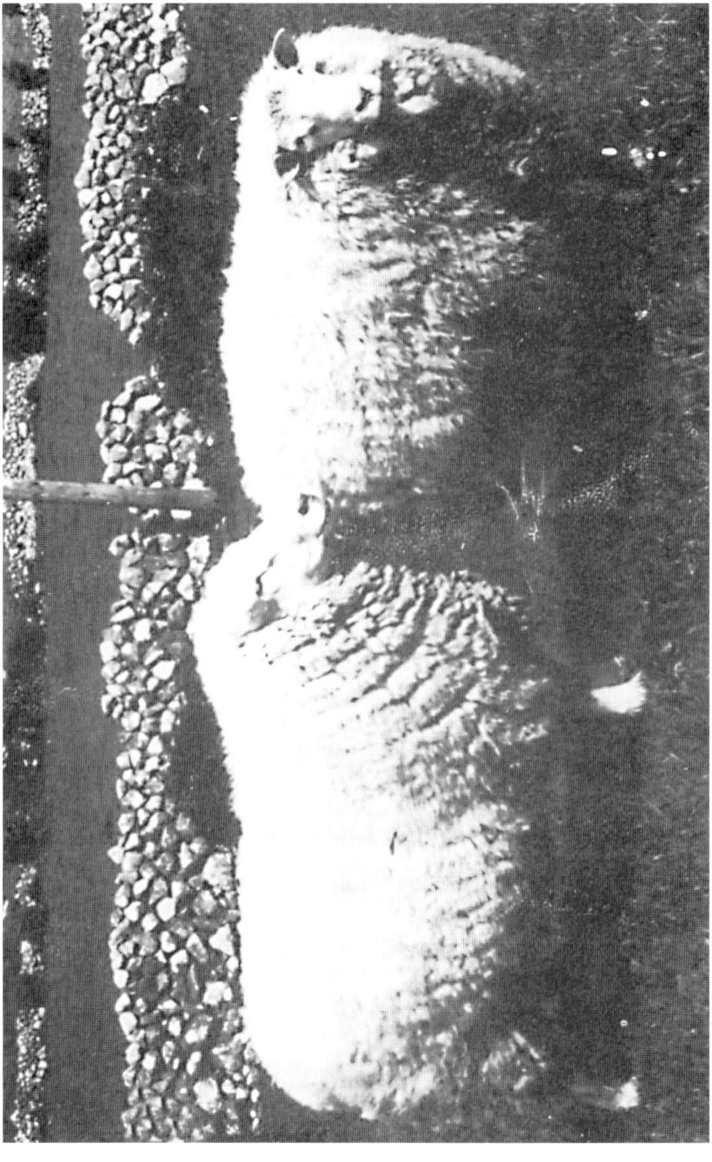

Two Galway sheep in typical setting, from Agriculture in County Galway, 1980, ACOT and Galway County Committee of Agriculture.

TABLE OF CONTENTS

Pen of five purebred Galway Ewe Lambs, 1st Prize winners at the Galway Sheep Breeders' Annual Show and Sale 1999. Presented by Tom Murphy, Shralea, Creagh, Ballinasloe. Sold for 65gns. each.

A typical Galway lamb

INTRODUCTION

Ireland's Old Breeds of Livestock:

The efforts which have been made, and which continue to be made, to increase the productivity of livestock reduce their genetic diversity. The prospect of cloning greatly increases that threat. Because the modern highly productive breeds give greatest output in intensive farming systems, there has been a reduction in total numbers within the more traditional breeds and an extinction of some breeds.

The expression "extinction is forever" continues to have validity with regard to the loss of rare breeds of domestic livestock. It is true that advances in DNA technology may enable ancient species to be recovered from portions of mummified tissue preserved by chance in unusual circumstances, but the application of such technology is most unlikely where rare breeds are concerned. The advantages of viewing living, breathing specimens - conserved through traditional farming methods on *terra firma* - is self-evident.

Even primitive (as opposed to traditional, and modern) breeds may be bearers of genes which could prove useful in changed circumstances. Thus, all livestock breeds - primitive, traditional and modern - are worthy of conservation. The Roscommon Long-wool breed of sheep probably is quite extinct at this stage, but may have conformed to our notion of 'traditional'. Its close-relative, the Galway, is another long-wool breed, but it continues to have a place in modern farming. The official descriptions of these two breeds reveal something of their similarities and reflect ancestral relationships. The Roscommon was larger, more long-legged and slower maturing than its surviving relative which has a better quality fleece. Unfortunately, the poor price commanded by wool at this time is just about adequate to pay for the cost of shearing.

Characteristics of Roscommon and Galway Sheep

Official Breed Descriptions:

[figures in the 'Roscommon' column refer to a percentage 'scale of points' from volume 1 of that Flock Book]

Roscommon	**Galway**
Head; [10] Face, white, long. Muzzle, strong in rams, more or less fine in ewes [a small quantity of wool on forehead not objected to] Ears, medium length, white or pinky, fine texture. Head gaily carried.	Head; Hornless; Back of head between and under the ears covered with short wool and showing a medium bob; Ears of good length and fine texture, and free from black/brown sports; Nostrils dark in colour and wide; Ram with bold masculine head of medium size.
Neck; [5] Moderate length, well set; in rams, stronger with a real good crest.	Neck: Thick, short, well set.
Shoulders; [15] Broad, oblique, well set in flesh fore and aft, and level with back.	Shoulders; Well laid, level and wide at top.
Chest; [10] Deep and wide	Chest; Wide, deep, well forward.
Ribs; [15] Long, well sprung, with a full flank.	Ribs; Well sprung; Barrell deep, thick, long with straight underline.
Back; [back + loin; 20] Long, level, well covered with meat and muscle.	Back; Long and level, backbone not prominent.

Loin; Broad, not too long; Tail broad, well set in.	Rump, Tail, Hind Quarters; Rump, wide and well fleshed; Hind quarters, deep, full, broad.
Legs; [legs + feet; 10] Straight, wide, with flat bone, woolled to knees and hocks, clean below; fore-legs well apart; hind-legs well filled with mutton.	Legs, Feet, Hooves; Medium long, straight, square, well apart, good bone and covered with wool to the knees/hocks. Slight feather on hind legs below hocks permitted but not desired. Hoof black, with sound horn.
Belly; [5] Well covered with wool [applies also to scrotum in ram].	Belly; Well covered with short thick wool.
Fleece; [5] Long, heavy; Staple broad, white of bright texture Hair or coarseness in quarters, etc. condemned.	Fleece; White, moderately long. Fibre, wavy and fine textured; Staple, broad, closely set and free from tips. Fleece in general as uniform as possible in fibre texture.
Skin; [5] Fine, soft, pink.	Skin; Light and pink
Carriage; [Head gaily carried]	Carriage; Gay and alert.

Galway Ewe with triplets (GCCA Report, 1973)

Reasons For Conservation:

(a) Cultural

In most cases, each of the primitive and each of the traditional breeds is associated with a particular country or province or district. They contribute to environmental diversity and counteract the monotony of uniformity. Some display distinctive territorial behaviour in their natural setting. They represent part of the heritage, history, the way-of-life and the landscape of an area.

This cultural dimension to these two breed categories becomes more important with the passage of time, and their continued existence becomes a resource for rural tourism. In some cases craft products (such as *báinín* knitwear) form part of the culture and the tourism.

(b) Scientific

Those of us who have lived through most of the 20th century are acutely aware of the extraordinary environmental change which has taken place - environment being the conditions or influences under which any person or thing lives or is developed. Despite all our technological advances, uncertainty remains as a major characteristic of the future. In changed circumstances some of the genes or gene combinations in primitive and in traditional breeds may prove useful to mankind, and the financial poverty of many civilizations keeps modern technology as a distant prospect. For this reason all such breeds should be conserved.

Furthermore, there are dimensions of science that impact on heritage. For example, the reconstruction of evolutionary pathways to demonstrate relationships between breeds can reveal affinities or connections between cultures.

(c) Agricultural

Some aspects of farming have been affected by vagaries of fashion. Subjective showring decisions, followed by inordinate publicity for the winning animals, gave rise to fashionable strains within modern breeds with a consequent erosion of less publicised strains. Merit classes based on objective measurements are a much better method for selecting superior animals and can be applied also to the less prominent breeds, but intense selection may narrow the genetic base.

Commercial livestock units frequently have successful enterprises based on a specific programme of cross-breeding, that is established breeds are used to produce vigorous cross-bred parents of the commercial end-product. A breed cannot continue in existence unless there is a central core or nucleus of pedigree stock which maintains the essential characteristics of the breed, which has become - through selection - "an artificial product" compared to the ancient ancestral race. Such intense selection pressures generate a move towards homogeneity within the breeds useful in commercial farming. The trend needs to be counteracted by the conservation of diversity within traditional sub-populations maintained to some degree outside of mainstream agriculture. The revenues to support such enterprises are generated in the tourist sector rather than in the food production sector. Heritage parks have an important part to play in this matter - a part which is not yet properly delineated and organised in this country.

The Purpose of this Book:

The aim is to draw attention to the need for this country to play its part in the global effort to stem the erosion of Earth's genetic diversity of livestock. A statement from the FAO[ii] that 30 percent of the 5,300 breeds (in 35 species) in existence today are in danger of extinction has little impact here. It is assumed generally that the statement applies elsewhere and does not concern Ireland. However, in 1990 there were less than 200 breeding females in the national herd of Kerry Cattle, and Dexter Cattle were almost totally extinct here. The Irish Moiled Cattle were revived through efforts largely centred in Northern Ireland. Roscommon Sheep appear to have become extinct in the period 1960 - 1990, and the Galway Sheep Breeders' Society experienced a survival crisis in the early 1990s and has just had another serious drop in membership in 2001. Thus, there are extinctions of native breeds, problems for their supporting bodies, and a loss of genetic diversity within surviving breeds. Relevant state departments need to have strategies for action and for cooperation with bodies such as the Irish Rare Breeds' Society and the Irish Genetic Resources Conservation Trust in conserving our native livestock heritage.

Lincoln Ram

Leicester Longwool (from leaflet produced by the Leicester Longwool Sheep Breeders' Association)

Ryeland Ram

One of the 1st prize pen of three Roscommon Lambs at Royal Dublin Society's Show, August 1895 at which date he scaled 14 stones (89 Kg), live weight. Bred by Mr. John Keane, Templepatrick, Mullingar.

— CHAPTER ONE —

IRISH LOWLAND SHEEP IN THE 19th CENTURY

Evidence from the Statistical Surveys:

A return to peaceful conditions in Ireland as we advance into the 21st century creates opportunities for cooperation between all sections of the island community in the conservation of biodiversity. The momentum towards such conservation is perhaps greater in Northern Ireland than in the Republic. Conditions in Ireland could scarcely have been described as peaceful in the 18th and 19th centuries when the establishment of breed societies gathered full pace, although some places and periods were more tranquil than others.

From about 1755 in Britain, Robert Bakewell of Dishley in Leicestershire, through selective breeding, began to bring about improvements in the qualities of horses, cattle, sheep and pigs. The principles which he applied so successfully with *Longhorn cattle* and *Leicester sheep* were taken up by John Ellman of Glynde with *South Down sheep* and by others to transform local races of domestic stock into distinct breeds. Thus, when Arthur Young came to tour Ireland in 1776 and in 1779 he noted that Mr. Gore of Ballina had excellent sheep that had been "improved by rams from England".

Good sources of information about 19th century breeds of sheep in Ireland are the 'Statistical Surveys of the Counties' [not all counties were covered] published at the beginning of the 19th century by the (Royal) Dublin Society.

Hely Dutton put together the Statistical surveys for Clare and Galway (1808). In the Clare survey he states (p.89): "Sheep have been greatly improved in their shape by the introduction of Leicestershire rams, but materially injured in the quality of their wool; this was formerly short and fine, adapted to the soil and manufactures of that part of the county, especially that produced in Burrin (Burren), and bore a high price at the Ballinasloe fair in July; since that period it has become coarser, and the old women regret the introduction of the *Dexters*[1] (as they call them),

[1] The italics are in the original text

which they say spoiled their wool ... Mr. Blood, of Riverston, who possesses large tracts of rocky soil in Burrin, has with great judgement introduced the breed of South Down Sheep, selected from the choice flock of the Marquis of Sligo and Mr Wynne of Hazelwood, convinced, that not only from being the finest woolled sheep in England, but also from their approximation in propensities and in appearance to the ancient breed of the country, they are to be preferred for this kind of soil.

I have often with great pleasure viewed his lambs, produced by the first cross of South Down with selected native ewes; the improvement was astonishing; there can be little doubt that their produce will be covered with wool nearly as fine as the South Down; I have a sample of this wool that is nearly equal to an South Down ... after that (the 4th cross) if care has been taken in selection no difference will be perceived. In a sheep-rearing country such as Burrin, possessing a short bite, and requiring activity to gather it, there can be little doubt of the inestimable value of the South Down breed of sheep, and that the breeders have sustained a heavy loss by the deterioration of their wool". (A footnote mentions sale of Blood's land and dispersal of South Downs out of the country and continues "A piece of fine broad-cloth manufactured from the wool of these sheep has obtained one of the premiums given lately (March 1808) by the Farming Society of Ireland, yet the breeders permitted this valuable breed to leave the country").

On pages 91 and 92 Dutton gives tables of sales of "clothing wool" at Mr Berry's, North Anne St., Dublin, in 1807 and includes prices. In his list he includes mention of people and their sheep breeds: Earl of Farmham (Spanish and Ryland); Nathaniel Trumbull, Esq, (Ryland); Thomas Trench, Esq (Spanish and Ryland); Rev. James Symes (Wicklow and South Down). The folk memory in Galway suggests that the Ryland was used at an early stage to help in the formation of the Galway breed.

The Merino Factory for Irish Superfine Cloths:

The reference to Spanish sheep in Dutton's list indicates that exotic breeds of sheep may have been involved to a small degree in the improvement of the native sheep of lowland Clare, Galway and Roscommon. The Spanish sheep probably came from the same source as those at the 'Merino Factory for Irish Superfine Cloths' established by Nowlan and Shaw near Stoneyford about seven miles from Kilkenny in 1810 [iii]. The enterprise was for the purpose of educating and employing the industrious poor and of encouraging the growth of fine wool in Ireland. The farm which formed part of the enterprise carried about 600 Merino sheep and the workers were apprenticed for four years before being [deemed] proficient in the production of woollen cloth.

The initial buildings were not completed until 1815 and, on these and on the machinery installed, a sum of 30,000 sovereigns was expended; a further 10,000 sovereigns on additional buildings and machinery was expended in 1818. During the progress of the works a certain number of apprentices were selected from the children of the neighbouring peasants, and instruction and moral discipline gradually introduced upon a plan calculated to establish and perpetuate the manufacture, and at the same time entitle it to the approbation of the legislator and moralist... Both males and females, from the very day of their engagement, receive weekly wages (payable every Tuesday to their parents)... The girls are (besides) instructed in the use of the needle and spinning wheel; nor are the boys permitted to be ignorant of the use of the reaping hook and spade... The Proprietors ... have provided that from the hour of rising til bedtime, their apprentices shall be employed, either usefully at business or innocently in active exercise conducive to health.

In 1818 there were over three-hundred employed and an appeal was made to the Noblemen and gentlemen of Ireland to purchase the produce in preference to that of other countries. An English Traveller, writing anonymously in the pages of the *Freeman's Journal* of November 14, 1817, included an account of the Merino Factory: ... "this factory, with the extensive farm which surrounds it, commences as it were, and completes within itself, and by its own powers, every process from the growing of the wool on the living carcass of the Merino sheep to the finishing of the piece or end , as it is technically called, of the best superfine cloth...".

The existence of the Factory and the mention of Spanish sheep in Dutton's list is evidence that Merino's were present in substantial numbers at that time. Later, the Toler-Aylwards[2] who were prominent in the establishment of the Roscommon Long-woolled Breed in 1895 had a Merino flock at Strokestown, Co. Roscommon, until about 1914.

[2] Personal communication.

Evidence from other sources:

It may seem strange to find that Andrew Corrigan [iv](1859) referred to the sheep of the Ballinasloe area as "Cotswolds", but sheep of the latter breed have resemblances to those on the limestone areas of the West of Ireland. In his view the sheep of the Ballinasloe area were of "a heavy constitution, (and for) weight of carcass, quality of fleece and superior wool, were never yet surpassed".

There is a brief reference to Roscommons in Fream's Fourteenth Edition of Youatt's The Complete Grazier [v] (1900):

"In Ireland the chief native breed is the Roscommon, which has been greatly improved during the past twenty-five or thirty years by the introduction of Leicester blood, and by careful selection. There was a limited but good show of these sheep at the Windsor Show of the Royal Agricultural Society in 1889". There is a further reference to statements by Fream in the Introduction to volume four of the Roscommon Flock Book (1899), including the statement that "It is a big-bodied sheep, covered with a long heavy silky fleece, its good qualities being due to careful crossing".

Clearly, there were Merino, Ryeland or Ryland (these came from Herefordshire[vi]), South Down, and Leicester Sheep in Ireland in the 19th century. It is probable that genes from all of these were selectively incorporated in the progenitors of Galways (The Galway Sheep Breeders' Society was founded in 1922). However, it would appear that only South Downs and Leicesters were the main outside contributors to the development of the Roscommon breed.

— CHAPTER TWO —

THE ROSCOMMON SHEEP BREEDERS' ASSOCIATION

Formation:

The improvements in farm livestock that progressed through the 19th century gradually extended throughout Britain and were not without influence on the livestock husbandry practised by some of the Anglo-Irish and native Irish. Thus, Major M. J. Balfe of South Park, Castlerea, Co. Roscommon, claimed in 1896 that the flock of *Roscommon sheep* on his estate originated with a flock established there in 1790[vii]. Likewise, Matthew Flanagan of Tomona, Tulsk, Co. Roscommon and F. A. Flanagan of Cartron, Mantua, Co. Roscommon, each claimed in 1896 that their large flocks of *Roscommon sheep* had been "handed down to me by my father, and to him by his father" (ibid). Many flocks of the breed were recorded in 1896 as having been established at intervals from 1840 (see List 1). However, dates of establishment of several flocks were not given in the 1896 publication (see Lists 2 and 3).

The rules of the *Roscommon Sheep Breeders' Association*, framed in compliance with the Friendly Societies Act (1875 and Amendment 1876), were printed in full (on pages v to xi) in volume one of the Roscommon Flock Book (1896). There are entries for a total of 57 registered flocks in that volume. Of that number, 16 consisted of 5 ewes each, or less, giving a sub-total of 51 ewes in all. A further 5 flocks each consisted of 6 ewes, and there was one flock of 9 ewes and one of 10 ewes, giving another sub-total of 49. The ewes in these smaller flocks were mated with rams from large flocks; 16 rams being used on the 100 ewes of the smaller flocks in the 1895-'96 season. Nine of these latter rams came from flocks owned by Flanagans (Mathew; F. A.; Patrick P.)

List 1: Flocks were established in Co. Roscommon by :

E. A. Stafford	Clogher, Mantua,	1840
C. & J. Cotton	Longford House, Castlerea,	c1842
Terence Glancy	Frenchlawn, Castlerea,	1856
Michael Feeney	Rathnagly, Tulsk,	1866
Bernard Rogers,	Ogula, Tulsk,	1866
James Rogers, Junr.	Ogula, Tulsk,	1866
Patrick Dufficy,	Carrowgarrow,	1869
Patrick J. Fitzmaurice,	Elphin,	1870
Patrick C. Morriss,	Frenchpark,	1870
Patrick Rorke	Cloonyquin, Tulsk	1870
Wm. Dolan,	Ballyconboy, Mantua, Castlerea	1871
Laurence Fahy,	Carton, Tulsk,	1871
Patrick M'Cormack,	Ogula,	1875
Bernard M'Cormack,	Rathynagly, Tulsk,	1876
George R. Acheson,	Rush Hill, Drumsna,	1876
John Coyne,	Ballyconboy, Mantua P.O.,	1876
Michael Connor,	Cloneiffer, Mantua,	1876
Timothy Rorke,	Comague, Tulsk,	1876
John O'Connor,	Rathcave, Tulsk,	1880
Edward Deffly,	Hill Street,	1886
Jno. Shanly,	Lugboy, Elphin,	1892

List 2: Flocks also were established in other counties:

Captain Thomas	Longworth Dames Greenhill, Edenderry, County Offaly.	1866
Matthew J. Kenny	Freagha Castle, Miltown Malbay, County Clare.	1870
H. J. C. Toler Aylward	Shankill Castle, Kilkenny	1871
A. H. Smith-Barry	Cordangan Manor, Tipperary.	1892
Pat & John Keane	Templepatrick, Moyvore, Mullingar.	1895
Patrick P. Flanagan	Rathtarmon, Frenchpark, Co. Sligo	Time not stated
Fred P. Smith	Kerit Castle, Crossderry, Co. Cavan.	Time not stated

The total number of ewes registered in the 1896 issue of the Flock Book was 3,557 and the number of rams used with these was 79, or an average of approximately 45 ewes to each ram. Slightly more than 200 ram-lambs born in 1895 were registered. Two rams born in 1891 continued to be used as sires in 1895, and 8 rams born in 1892 continued to be so used.

People prominent in Irish life gave support either through ownership of pedigree sheep and membership of the Society (The O'Connor Don of Clonalis, Castlerea) or through membership only (Count Arthur Moore of Mooresfort, Co. Tipperary, and members of the Ganly family of livestock auctioneers, Usher's Quay, Dublin).

Three Roscommon rams, 1906 (CMA, Vol. 10)

Roscommon ram, Adam (CMA, Vol. 10)

Roscommon ram, Cup-Bearer, 1896 (RFB, vol. 2)

Roscommon shearling ram, 1898 (RFB vol. 4)

Two-shear Roscommon Ram, 1901 (RFB, vol. 6)

Three Roscommon ewes, 1896 (RFB vol. 1)

Subsequent volumes of the Flock Book confirm that the number of flock owners continued to grow during the first decade of operation of the Association (List 4).

The eight signatories to the rules of the *Association* on the 17th July 1895 were The O'Connor Don; Right Hon. A. Baron De Freyne; Major M. J. Balfe; John L. Cotton; Luke J. Irwin; Joseph Banahan; Timothy Rorke; and Mathew Flanagan. The latter was listed as "Secretary", the others as "Members".

The Registered Office of the Society was stated to be at Tomono, Tulsk, Co. Roscommon.

In witness to the rules there was an *Acknowledgment of Registry of a Society under a limited Special Authority* dated 30th August, 1895 from Daniel O'C. Miley, Assistant Registrar (of Friendly Societies), Ireland. A copy of the rules was kept at 16, Dame Street, Dublin.

More than sixty years later, the Report of the *Roscommon County Committee of Agriculture* (1958, page 56) noted that "Roscommon sheep are recognised to be one of the oldest native breeds even though critics may differ when discussing their finer points, but very few will fail to identify this breed by its great size, strength of bone and quality of wool".

List 3: The remaining flocks recorded in volume one of the Flock Book and established in Co. Roscommon at times not stated were:

Joseph Banahan Renny, Tulsk,
Michael Broderick, Castlerea,
[Thomas Croghan, Carrowgarve, “given to me by my father”]
Patrick Cunnane, Lisnagard, Elphin,
Patrick Connor, Ballyconboy, Mantua,
T. and P. Conry, Raheen, Elphin,
Denis Daly, Cloonieffer, Mantua,
James Dolan, Ballyconboy, Mantua,
Patrick Dolan, Ballyconboy, Mantua,
John Durr, Mantua, Castlerea,
John Woulfe Flanagan, Lecarrow, Boyle,
Thomas A. Flannery, Ballaghadereen,
John Flanagan, Ballyconboy, Mantua,
Timothy Higgins, Cloonieffer, Mantua,
Luke J. Irwin, Raheen, Mantua,
Michael Kelly, Castlerea,
[Michael Kelly, Rathcroghan, “Got them from my father”]
Terence J. M’Loughlin, St. Patrick’s St., Castlerea,
Michael Mitchell, Cloonfinlough, Strokestown,
The O’Connor Don, Clonalis, Castlerea,
John O’Farrell, Strokestown,
John J, Ryan, Hollywell, Roscommon,
Michael Rorke, Lisadurne, Elphin,
Patrick P. J. Taaffe, Rathcroghan, Tulsk,
Thomas and C. White, Rockfield, Roscommon,

List 4: Flocks given in later volumes of Flock Book:

James Acheson, Aughlougher, Croghan,
Charles P. Adamson, Lacken, Glasson, Athlone.
Rt. Hon. Lord Barrymore, Cordangan Manor, Tipperary,
Wm. Carlos, Lisheen, Castleplunket,
Francis Cunnane, Flaska, Tulsk,
John Cunnane, Flaska, Tulsk,
Ed. P. Daly, Mayne, Kilmeedy, Co. Limerick,
Francis Darcy, Rathnagley, Tulsk,
Edward Deffly, Hill Street,
Wm. Dolan, Ballyconboy, Mantua,
James Dolan, Ballyconboy, Mantua,
Patrick Doorly, Clooneebane,
Rt. Hon. Lord de Freyne, French Park
John Egan, Ballymurray, Mantua,
Francis J. Flanagan, Tomona, Tulsk,
Michael Farrell, Flaska, Tulsk,
Patrick Keane, Castleplunket,
Patrick Keane, Junr., Castleplunket,
Bernard McCormick, Rathnagley, Tulsk,
Thomas A. Mapother, Kilteeven,
Wm. Neary, Moat House, Creggs, Co. Galway
Michael F. Neary, Ahalative House, Creggs, Co. Galway
O.W. O'Grady Young, Harristown Castle, Castlerea,
Capt. Hon. Otway Cuffe, Sheestown Lodge, Kilkenny,
Patrick Rogers, Rathnagley, Tulsk,
John Robinson, Lisnaglassick, Longford,
J. J. Ryan, Hollywell,
Edward Scott, Flaska, Cloonyquin,

Some Success:

"In 1897 at the RDS Show a special prize of £5 was offered for the best group of 15 sheep of one breed in the long-woolled classes. Four judges officiated and three of them were unanimous in giving the prize to the Roscommons, thus leaving the Lincolns, Leicesters and Border Leicesters in the shade" (Preface to volume 3 of the Flock Book, 1898). The prominence given to sheep husbandry through publication of the Roscommon Flock Book is in sharp contrast to the lack of interest in sheep of some contemporary authorities on the practices of farming. Edmund Murphy, who became the first Professor of Agriculture at the Queen's College Cork, made no mention whatsoever of sheep in his small book[viii] on agriculture which was published in 1849, with a second edition in 1853.

In the Preface (page xv) of the Flock Book (1896) it is stated that

> All sheep offered for entry have been carefully examined, and inspected, and in the right ear of those accepted, a ♣ shamrock mark will be found perforated. While seeking information as to their origin, the Association used every means within their reach and have been informed on all sides, that the long-wooled white faced Roscommon Sheep are the native breed of Ireland.
>
> And it is a well-known fact that in Roscommon, some flocks have been handed down for generations from father to son, and all those years bred true to type and colour, so that Roscommon Sheep can claim pedigree, not by giving three dams as other animals do, but by the enumerations of men.

The following "advertisement" which appeared in volume six of the Flock Book reinforces other evidence that members of the Association made efforts to promote the Roscommon sheep breed in several countries.

THE AGRICULTURAL SOCIETY OF ARGENTINE,
in their
THIRD INTERNATIONAL EXHIBITION OF CATTLE,
have awarded to
Messers. W. COOPER and NEPHEWS
a Nickel Medal, as Second Prize,
for a lot of five SHEEP OF ROSCOMMON BREED, presented
in the Sixth Class.

BUENOS AIRES
Septiembre 8 de 1898

Proprietarios:-

Senor Don M. J. Balfe. Senor Don P.J. Fitzmaurice
Senor Don T. Rorke. Senor Don G.J. Cotton (2)
Senor Don F. A. Flanagan (2) Senor Don G. R. Acheson
Senor Don M. Flanagan

Julian Frers, Presidente.
Jose Mariadetchával, *Secretario*

However, a sombre note as a closing paragraph to the *Introduction* to volume six of the Flock Book, 1901, (page xxiii) reminds us of the extreme difficulties faced by livestock farmers in the Meigh, Co. Armagh and Cooley districts of Co. Louth a century later (Spring 2001) due to an outbreak of foot-and-mouth disease in that area.

The 1901 note read as follows: "If it were not for the closing of the ports of the Argentine Republic there is little doubt but that we should have done a satisfactory trade in exports (of Roscommon sheep). But with the total disappearance of foot-and-mouth disease in England there now appears a fair prospect of the expected export demand being at last experienced".

Three Roscommon ewes, 1898. (RFB, Vol. 4)

Michael Conneely, Monivea, Co. Galway. Champion aged Galway Ram 1997 – Gurteen Agricultural College. RDS Class.

— CHAPTER THREE —

THE INFLUENCE OF BRITISH BREEDS

Leicester and Lincoln Influences:

Thomas Baldwin, Superintendent of the Agricultural Department of National Education, and Director of the Glasnevin Institute (Albert College), wrote [ix] in 1881 about the work of Robert Bakewell of Dishley Grange in Leicestershire in improving the Leicester breed of sheep and went on to say:

> The Leicester is the best breed for improving the sheep of Ireland. In peculiar situations, other breeds might answer better; but we think as highly of the Leicester for improving our sheep as of the short-horn for improving our cattle. There is the most ample practical experience to support this view. The fine sheep which are met with in many parts of Ireland under the name of *Ballinasloe sheep*, have been produced by crossing the native sheep with Leicester rams. The most skilful and enterprising flock masters in the country regularly use Leicester rams, and by doing so have enriched themselves and benefitted the community. We desire to see this example followed by the small farmers. It is true that a small farmer cannot afford to buy or hire a pure-bred Leicester ram; but by clubbing together, a number of small farmers could make up the five or six guineas for which a pure-bred animal of fair quality can be had.

A brief account of Roscommon Sheep was published in the *Standard Cyclopedia of Modern Agriculture* [3] (1910, volume 10, pages 141 - 143[x]). The account was prepared by O. W. H. Roulston, B. A., Inspector of Statistics to the Department of Agriculture for Ireland; Ex-Instructor in Agriculture (County Cork Committee); some time Editor of the *Farmers' Gazette*, and author of "Flax Growing in Ireland". Roulston graduated as

[3] Full title: The Standard Cyclopedia of Modern Agriculture and Rural Economy, 1910, (12 volumes) edited by R.P. Wright, The Gresham Publishing Company, 34/35 Southampton Street, Strand, London.

an Associate in Agriculture of the Royal College of Science in 1903. His account reads as follows:

This breed of large, long-wooled sheep takes its name from the county where it was first most widely established. Nothing very definite can be stated about the origin of the breed, nor of the types employed to fix the outstanding characteristics it at present exhibits. The claim made on its behalf to be purely native, i.e. to have been evolved by careful selection on the part of breeders in the counties where it has long found most favour, must bear considerable modification. There is a suggestion that the great size and frame of the typical Roscommon is due to the introduction into certain western counties of Ireland during the 17th century of big English midland sheep of a type then possessing much in common with the Lincoln[xi] breed as we now know it. However this may be, there are without question striking features of resemblance in the modern Roscommon which serve to substantiate the belief that the resultant breed-type is due to such a cross with the original Irish native sheep in these districts. When we come to nearer the close of the 18th century we reach surer ground, inasmuch as definite proof exists that Leicester rams, then brought into prominence by Bakewell's results, were employed to influence the breed characteristics. To what extent this introduction of Leicester blood modified the type cannot be definitely decided, but there can be little doubt that increased quality and symmetry as well as possibly more proclivity to early maturity resulted therefrom. There is some belief that the Leicester influence diminished size, but this does not appear to have been permanent; possibly the class of close-feeding limestone soil which prevails over the midland area long devoted to the breed had no small effect in maintaining the frame and substance for which the Roscommon sheep is noted. Similar results may be observed in the big upstanding large-framed cattle which are raised on the pastures over the same area.

During the last half-century the breed has continued to take a still more increasingly prominent place in the classes of Irish sheep. This result has been in large measure due to the fostering care of the chief breeders throughout Roscommon and the adjoining counties. In 1895 a decided step to further the interests of the breed was taken by the establishment of the Roscommon Sheep Breeders' Association. Among its other objects this Society which includes all the leading adherents of the breed in Ireland, devotes itself to the breeding of Roscommon sheep at home and abroad, the publication of an annual flock book, and the periodical compilation of transactions associated with the breed.

Galway, Mayo and Roscommon are essentially the home of the breed, though it is also to be found eastward through Westmeath and over much of the midland area, as well as extending southward over Tipperary and Limerick, and in some of the grazing and tillage districts of the north. In Galway, Mayo and Roscommon, however, it has the strongest preponderance, more especially as these are three of the most extensive sheep-rearing counties in Ireland. In 1908 the total Irish sheep stock aggregated 4,126,106 of which 1,181,198 were returned from Connaught; and of this number again the three counties in question were reponsible for no less than 1,096,373 ...

The points of the Roscommon sheep cannot be more clearly and explicitly stated than by the following abbreviation taken from the scale laid down in the flock book;

[see conformation details set out above in the "Introduction"pp. 14-15]

Apart from its (con)formation the Roscommon sheep is possessed of a very vigorous constitution, which enables it to thrive readily on the bare, flat lands, which form its chief run. Though grazed largely on such runs, which are possessed of little shelter, its innate hardiness renders it free from the constitutional troubles that exposure frequently brings to sheep of other breeds. Further, it is an extremely active feeder, and its particular soundness of feet enables it to range widely when grazing.. The ewes are good milkers, and on this account, as well as of its size, the breed stands in high favour for crossing purposes. The Roscommon and its different crosses enter largely into the Irish ewe breeding stock of over 1.5 millions. Roscommon ewes are very successfully crossed with Shrop and Oxford Down rams. The result affords lambs which, although not so early maturing as those of other breeds, find much favour with butchers; many of these lambs, however, are kept over and sold as hoggets in the following year, when they fetch very satisfactory prices, and enter largely into the supply of the London, Manchester, and Liverpool markets.

Some very heavy weights of Roscommon sheep have been recorded. It is not uncommon for lambs to average 28 lb. per quarter, and shearling ewes and wethers have when dressed been found to average 37.5lb. (16Kg)

per quarter. A leading authority has put the average weight of dressed carcass from two-shear wether Roscommon sheep at 100 lb. Roscommon mutton is of excellent quality, and shows a well-mixed even grain, due no doubt to the active nature of the breed and their tendency to muscular development. ... The average weight of fleece clip may be put at 10 lb
The chief flocks of pure-bred stock, to mention but a few - are those kept by Mr. Mathew Flanagan, Tulsk, Co. Roscommon (Secretary to the Roscommon Sheep Breeders' Association); Major M. J. Balfe, Southpark, Castlerea; Mr. George R. Acheson, Drumsna; Mr. John Keane, Baltacken, Moyvore; Mr. P. J. Fitzmaurice, Elphin, &c.

As Roscommon sheep cannot compete with other breeds for early maturity, very few lambs are put on the market till the middle or end of August. Indeed throughout the chief breeding districts the great bulk of the lambs are not exposed for sale until the large autumn fairs held during September and October. The most important of these fixtures are the September and October fairs of Ballinasloe, the September fair at Banagher, and the October fair of Tuam. ... Rams from leading pure-bred flocks are also largely sold here, and the prices range from £6 to £10, according to demand and the prospects of the sheep trade. A large proportion of the Roscommon lambs and wethers offered at Ballinasloe are bought by graziers in the richer districts of the south and midlands, who run them over to be disposed of when fat during the following spring and summer. A fair supply of Roscommons and their crosses, when finished, find their way to the Dublin market; but as large sheep are not in favour with Dublin victuallers, the bulk of fat Roscommon sheep are shipped across channel by exporters, and disposed of in such centres as Liverpool and Manchester.

Views of Major Balfe on Roscommons:
The *Cyclopedia* article quoted Major Balfe of Castlerea:
For early maturity, lambs bought in this county for £2 last September and October fetched up to £4 in the Dublin market after Christmas. As the quality of mutton cannot be surpassed, hoggets go to supply the London markets. Kept on natural fare without pampering, they preserve their constitutional vigour and retain their great size and beauty of formation, and mature early. The ewes are very prolific, and good nurses: hardy and stand the winter without shelter; active in the pursuit of food; are good

growers and weighers when they reach the abattoir, and possess a heavy fleece of long silky wool. The ewes are splendid for crossing with a Shrop ram - for market lambs; and [Roscommon] ewes command the highest prices at all the principal sales and fairs. They produce the biggest weighers and are fair milkers. [Roscommon] lambs are very hardy and do without any artificial feeding in the winter.

Not long since, I sold thirty hogget rams at a big figure to a Buenos Ayres buyer, and they were well pleased with them in that country. I have also sent sheep to Russia; and I am sure in the near future there will be a run on this breed by all foreigners, as no sheep come up to the weight of them and the good long wool that they produce, together with their hardiness of constitution and the good quality of their mutton, and their commanding appearance even if sold as stores.

Views of Mr. Keane of Moyvore:

The *Cyclopedia* quoted Mr. Keane of Moyvore as follows:

Roscommon sheep are suited for any land of medium quality, but they grow and fatten earlier on the rich lands of Meath, Kildare, Westmeath and Roscommon; and animals of this breed sent to Russia, New Zealand[4] and the Argentine Republic have proved very successful. Dublin and London are the principal markets suited for these sheep. I may give an instance of a transaction I made last December. I sold nine Roscommon shearling ewes and wethers to a Dublin firm of victuallers; when dressed they made the enormous average of 37.5 lb. per quarter. There was very little waste or fat on the carcass when the weight of the sheep was considered. [Roscommons] are not so early for market as Shropshire crosses, but the lambs grow bigger and make higher prices in the fat and store markets during the months of July, August, and September. ... The breed is especially suitable for crossing with Oxford Downs ... and will live and thrive in the most unfavourable climatic conditions.

A pamphlet outlining the general characteristics of Irish agriculture was issued by the Department of Agriculture in 1927[xii]. It had little to say about sheep but included a photograph of "A Prizewinning Roscommon Ram A Pure Irish Breed".

---

[4] Exhibited by Mr. John Davies of Heatherly, Kereru, at Manawatu Show, see Introduction to volume 10 of the Flock Book (1906)

The State's Official Handbook [xiii] published in 1932 restricted mention of sheep to a twelve-line paragraph in a nine-page article on agriculture by Joseph Hanly.

It stated that the principal breeds favoured are the Roscommon, a large white native sheep well suited to the midlands; and the Galway sheep, a white, compact animal of medium size, very hardy and prolific, recently established. This latter type of sheep is almost universal in Galway and Clare. In addition to these, Oxford Downs, Shropshires and several other lowland breeds are favoured locally.

All of these sources point to the existence over a long period of time of large-bodied, white, hornless sheep on the lowland grasslands of Roscommon and Galway, or as expressed in 1958[xiv]:

Roscommon sheep are recognised to be one of the oldest native breeds even though critics may differ when discussing their finer points, but very few will fail to identify this breed by its great size, strength of bone and quality of wool.

Nicholas Cotter - a supporter of Roscommons:

A paper [xv] by Nicholas P. Cotter (1939) also referred to the use of Leicester rams by some of the "landed gentry" to bring about improvements in the native white sheep of Roscommon. He added that the general consensus of opinions was "that the original breed which predominated in the county 130 years ago was a heavy, coarse-grained, long-wooled type of sheep which produced mutton of little flavour and no delicacy. The carcase was long, flat-sided and with large bones set on thick, rough legs. The animals were slow feeders, and were not sent to market until over three years of age. The fleece was heavy and rather coarse in quality".

Cotter's view on the use of Leicesters is not substantiated by statements in the Roscommon Flock Book, volume 1, (1896). In fact on page xxi of the Introduction (pages xix-xxv) it is stated unequivocally that:

> "It is held by some authorities that the present breed is the result of crossings with English blood, but the general consensus amongst old flockowners in the county is that they are the native breed, improved by judicious blending of the various predominant qualities of the sire with the Flock that is, by taking advantage of the valuable

> peculiarities encouraging their development, and by degrees rendering them more permanent. It is an old saying that the selection of one good tup may be the making of a whole Flock, and in this connection it would be hard to overestimate the care and judgement that some of the best-known breeders of the present day bestow in inspecting flocks before they select a change of blood in the purchase of a young sire or two every season".

In contrast, the Introduction to volume four of the Flock Book (1899) stated what appears to be a more correct view: "The superior qualities which the infusion of a little Leicester blood, imparted to the breed at its very initiation, have been developed to the utmost limit by those business-like admirers of the type who brought it to its present state of perfection; and, as recent records have shown,, their efforts have not been altogether unavailing. Now we can claim as our own a sheep which has shown itself, in the midst of the strongest competition, to be much above the average".

Views of another sheep expert (Mr. Pat Boland):
Some decades ago, residential students of the Albert College in Glasnevin, and students of the Faculty of Agriculture there, experienced the skills and wealth of practical knowledge of sheep displayed by Mr. Pat Boland of the farm staff of the College over a lengthy period. His article on "Sheep"[xvi] (1960) illustrates his depth of knowledge. He noted that "with the exception of a few farmers in one or two of the southern counties arable (sheep-)farming is not done nowadays ...
The Down breeds are more suited for close folding on arable land and are noted for the high quality of wool and mutton produced. The most popular of these are the Suffolk, Oxford, Hampshire, Southdown, Shropshire and Ryland, all of which we have represented in Ireland, with the exception of the Southdown, which I have never come across here ...
The long-wools are chiefly represented by Roscommon, Galway, Lincoln, Leicester and Border Leicester ... The long-wools are not noted for good mutton although the Galway is an exception, but produce large clips of wool of high lustre and there have been instances where prize Roscommon rams clipped 24 and 25 lbs of wool at first shearing ... The origin of the breed dates back to the 17th century and it is supposed that the size

originated through the introduction of the Lincoln breed over one hundred years ago. Leicester rams were also used fifty years prior to this to influence the breed characteristics ... (The) flock book has been issued annually ever since 1895, until some years ago. Now each breeder keeps his own records.

It is well known that the present Galway sheep breeders purchased Roscommon rams from the leading Roscommon breeders up to forty years ago. These rams were mated to Galway ewes and the resultant progeny, being carefully selected, has led to the establishment of the present Galway sheep of today ... Time after time I have been asked the difference between these two breeds of sheep, but I find it very hard to describe. The Roscommon has a finer and cleaner cut head and face with a very pronounced tuft of wool on the forehead [5]. (Of course the Galway may also have a tuft). The Roscommon is not so broad at the muzzle as its rival, it is slightly taller but otherwise identical.

When given suitable rams the Galway is the best of our native breeds for the supply of early lambs ... They are early to come to the ram, exceptionally good milkers, excellent nurses and very prolific, and not so susceptible to disease as the Roscommon. The soundness of their feet is a striking characteristic of the breed and it is attributed by some to the fact, that before the advent of trains, these sheep had to be driven long distances by road to Ballinasloe Fair, from whence again they were driven to various parts of the east and midlands. Naturally weakling fell out by the way and a sound-footed type was evolved. The wool is of a very high standard - 7 to 10 lbs per head can be taken as an average clip and commands the best price of the long-wooled group ... with the tendency of the modern housewife in favour of small joints, I think the Suffolk (x Galway cross) is best suited to her requirements ... When buying a commercial flock of these ewes the best specimens of the breed can be had around Tuam or in the little village of Headfort in Galway ... The Galway raised in the area between Craughwell, Claregalway and Athenry - this triangular portion of County Galway - produces the best wool we have, apart from the few pure-bred representatives of the Down breeds (Suffolk, Shropshire and Oxford).

[5] Authorities differ on this point. There is variation within both breeds in the amount of wool on the forehead. In official breed descriptions a tuft, or bob, was considered necessary in the Galway and "optional" in the Roscommon.

The LEICESTER LONGWOOL BREED

Not surprisingly, this white breed is native to the midland counties of England, although greater numbers were found for many decades in the Yorkshire Dales. The breed is considered to be dual purpose for meat and wool production and to be very tolerant of harsh and wet weather conditions. It is claimed that the birth survival ratio is higher for the Leicester than for any other pure breed. The number of pure bred Leicesters dropped to 350 in the 1970s and a plan of conservation became necessary. Description[xvii]: Mature rams weigh around 150 kgs and mature ewes up to 100 kgs. A fleece of 5 to 7 kilos of fine lustre wool is produced. The head is bold and strong on a short thick neck. The crown is well covered with wool, and the face covered with white hair. The ears are blue, fine and fairly long with occassional black spots. The muzzle is strong with even jaws, and nostrils dark in colour. The carcase is deep and of considerable length with full flanks. The back is broad and level and the ribs well sprung. The legs of mutton are wealthy with flesh coming well down to the hocks. The legs are of medium length and good bone well set up on pasterns on dark feet. The legs are covered with white short hairs or wool.

An early description by "Mr. Marshall" of the improved Leicester breed is printed as Essay XX (pp.386 - 404) in Hunter's Georgical Essays,[xviii] volume 4, 1803. It stated inter alia: "In the work of culling ... the grazier's guide is principally age. The ram-breeder, on the contrary, goes chiefly by merit: a ewe that has brought him a good ram or two, is continued in the flock, so long as she will breed: there are instances of ewes having been prolific to the tenth or twelfth year; but, in general, ewes of this breed go off at six or seven shear".

Galway ewe at pasture near Tara, April 2000

Tom Murphy with his 1st (MMM19, purchased by Fintan Connolly for 300 gns) and 2nd (MMM30, purchased by Leo Curran for 200 gns) prizewinning ram lambs at Athenry, 1996.

Michael Maher, Monivea (left) with Jimmy Courtney (Kilconly, Tuam) to whom he sold this hogget ram at the 1997 GSBS Show and Sale

Martin Maher (left) receives his trophy from judge P.J. O'Dea (Esker) for this 1st prize shearling ewe at Monivea Show, 2000

Martin Fallon (left) with his champion hogget ram, being congratulated by Seamus Cuddy, Ulster Bank (sponsor) 1997 GSBS Show and Sale

Best wool class, 1997, GSBS Show and Sale. John Maher, Monivea (left) holding JMM5) with Seamus Cuddy Ulster Bank (sponsor) and Michael Conneely (right) holding MCM9

Winners in the ram lamb class at the GSBS Show and Sale 1997. From left: Michael Conneely (Monivea), Joe Healy (Athenry), Seamus Cuddy (Ulster Bank) and Tom Murphy with the latter's champion (MMNI)

— CHAPTER FOUR —

FORMATION OF THE GALWAY SHEEP BREEDERS' SOCIETY

A Committee formed, 1922:

The meeting giving rise to the Galway Sheep Breeders' Society was held on the fifth of November 1922 [xix]. Christy Kerin [xx] of Oranmore presided over a further meeting of the Society at the Agricultural Station (now Mellowes' Agricultural College) Athenry, on 20/03/'23, with J. O'C. Sullivan as Honorary Secretary [xxi]. Others present included, Rev. Bro. Jarlath, W. Prendergast, J. Giveney, T. Cotter, and L. Ledwith. Arrangements were made for a number of ewe inspections; local agricultural societies were encouraged to provide prizes at shows, and a cup was to be presented to the Royal Dublin Society for award in Galway Sheep Classes at its Spring Show.

At the first AGM in June 1923 [xxii] Martin Farrell was elected Vice-President. Other members were T. Ruane, B. Taylor, G. Kelly, A. O'Dea., Willie Fahy (Moyvilla) and Pat Sice (Boyhill). A proposal from F. Quinn (Duras) and A. O'Dea (Belclare) that the Officers and Council from the previous year be re-elected was adopted. A grant of £25 from the Department was acknowledged. During the course of the meeting the natural high quality of Galway wool was emphasised.

The Sheep Breeding Scheme of the Department of Agriculture influenced the development of both Roscommon and Galway sheep. The Department's Report for 1923 noted as follows:

> During the period under review, certain County Committees set aside funds for the purpose of locating high grade pedigree rams with poorer farmers in the hilly districts. The Counties adopting this scheme and the types of animals located were as follows:-
>
> Galway (Galway Sheep), Roscommon (Roscommon Sheep), Waterford (Blackfaced) and Wicklow (Cheviot).

Two breed organisations in operation:

By 1924 the Roscommon Sheep Breeders' Association had lost much of its strength. The Flock Book for that year reveals that only five breeders registered their rams, and 187 ewes were admitted by inspection in an effort to spark a revival of interest.

For some time the two sheep breeders' organisations continued to operate side-by-side. Thus, in 1926 "as a result of inspections carried out during the year, 43 rams and 176 ewes were earmarked and registered by the Galway Sheep Breeders' Society and 38 rams and 224 ewes by the Roscommon Sheep Breeders' Association".

The Department's Report for 1927 noted that "In addition to keeping the Flock Book this (Galway) Society holds a local Show at which prizes are offered for the best ewes and lambs of the Galway Breed. The Roscommon Sheep Breeders' Association pursues similar activities in respect of Roscommon Sheep".

From Department of Agriculture Reports it appears that the numbers of pedigree rams located by County Committees of Agriculture in the first three years of the scheme were as follows: [zero in 1924], 11 in 1925, 7 in 1926, and 33 in 1927. Throughout the 1930s the committees of agriculture continued to foster the development of Irish breeds of sheep by means of grants out of their funds in aid of the work of societies devoted to the preservation of native types. Special assistance of this nature was afforded to the Galway Sheep Breeders' Society [xxiii]. In 1938 the Society reported [xxiv] that "the distribution of registered rams is effecting a very considerable improvement in the general quality of Galway type of sheep throughout a great part of the county". Mr. F. B. Higgins was president and Mr. T. J. Crowe secretary at the time.

Christy Kerin of Oranmore continued to have a major influence on the development of the Galway breed throughout the 1940s. He was a consistent producer of prize-winning sheep. The rams of outstanding conformation which he selected for exhibiting at shows were invariably single progeny of quality ewes. These selected rams he confined to a special paddock adjacent to the village and gave them supplementary trough feeding during the winter. His selection policy was guided by the skills of his flock manager - Mick Higgins of Moneymore townland - and contributed substantially to the definition of "type" in the Galway breed, but it led to a decrease in the fecundity of the breed as time progressed.

The agricultural adviser, Nicholas Cotter, had displayed a great interest in Roscommon sheep. However, without a conservation policy, if a breed does not find favour with changing consumer needs and tastes it is destined to go into decline with consequent loss of blood-lines. In fact, by the end of the 1950s the focus in sheep farming in the lowlands of the Western Counties had shifted entirely from Roscommons to Galways. A breeding project with Galway sheep was established by An Foras Talúntais in August 1963 with emphasis in ewes on prolificacy, mothering ability, and yield and quality of wool, and with emphasis in lambs on growth rate and carcase quality [xxv]. The Committee of Agriculture in County Galway commented on such developments in 1964:

> "In order to foster and develop the native Galway type of sheep, the Galway Sheep Breeders' Society was formed in November 1922. This body has received financial and other assistance from the Department and from the Galway County Committee of Agriculture. The Department's Officers have carried out inspections of a large number of sheep and selected from these for foundation stock some 600 ewes and 20 rams, which have been earmarked and passed as eligible for entry in volume 1 of the Flock Book of the Society, The selection so drafted, together with animals to be added as a result of further inspections, *should form a useful nucleus from which to build up a most valuable breed of fine-woolled, compact, early-maturing sheep, eminently suited for Galway and other adjoining counties (the italic lettering is inserted here for emphasis).*
>
> The Roscommon Sheep Breeders' Association has received similar assistance from the Department and from the Roscommon County Committee of Agriculture in respect of inspectorial and propaganda work.

The italicised portion portrays clearly the opinions of Department of Agriculture officials of the time. The final paragraph may be interpreted as a gentle withdrawal of support, in the early 1960s, for sheep of Roscommon type. The notion of conservation of native livestock was a weak concept in Ireland at the time.

In fact, before Cotter's death in June 1959 grants were awarded by the Roscommon County Committee of Agriculture for the improvement of sheep flocks in the County through the purchase by farmers, not of Roscommon, but of pure-bred Galway rams. Cotter's paper (1939) was reprinted in the 1969 *Annual Report of the Roscommon County Committee of Agriculture*, an action which suggests that sheep of Roscommon type continued to be important at that time in the County. The reprint was supplemented with advice derived mainly from research carried out by *An Foras Talúntais* on sheep farming systems . Scrutiny of the terminology used in the supplementary information gives an insight into the changing breed structure in Roscommon sheep-farming at that time. It was noted that:

> "local demand for lamb, including the increasing tourist trade in Western areas, absorbs quite a lot of the progeny. Large carcases do not meet with disapproval for this trade. In recent years factory buying has taken on in the area and the demand for smaller and better finished carcases is making the Suffolk cross more popular than the Oxford Down cross lamb".
>
> Two systems of sheep production were dealt with briefly: The All White System and the Cross-Bred System. It was estimated that 20 — 25 per cent of the sheep in South Roscommon are managed on the "All White" system which "entails the breeding of Galway-Roscommon sheep more or less in the pure-bred form". It also was noted that "the sales each year consist of (a) wether lambs and (b) draft ewes usually after the third crop".

Galways replace Roscommons:

Thus, the agricultural advisory service in Roscommon no longer referred to the main sheep-stock as *Roscommons* but as *Galway-Roscommons* (page 54) or as Galways (page 56). Likewise, the advisory service no longer spoke about the Ro*scommon Sheep Breeders' Association* but of "the recently organised *Roscommon Sheep Owners' Association*".

What happened in the interval 1939-1969 to bring about these changes in terminology and in organisation? The answer to that question centres upon the establishment of the Galway Sheep Breeders' Society and the good marketability of their product. Reasons for the increasing popularity of Galways and the accelerated decline of Roscommons may be gleaned from the *Annual Report of the Galway County Committee of Agriculture* for 1958. The section of the Report under the heading Sheep Breeding Scheme is reproduced below.

"In 1949, a scheme of grants towards the purchase of selected Registered Galway rams was inaugurated by the Committee. The scheme proved an immediate success. In 1958 a sum of £275 was allocated by the Committee to provide 55 grants of £5 each. The following are the principal conditions applying to the award of these grants:

(1) Inspection of rams for premiums under the scheme will be carried out at approved Shows and Sales of the Galway Sheep Breeders' Society, the arrangements for the inspection to be subject to approval by the Minister for Agriculture, whose Inspectors select the rams for provisional award of premiums.

(2) Only tattooed registered shearling (hogget) rams shall be eligible for inspection under the Scheme. Grants shall be awarded only in respect of rams inspected and passed at approved Shows and Sales of the Galway Sheep Breeders' Society and purchased by auction, at one of these sales, by a selected applicant.

(3) An applicant selected to purchase a premium Galway ram shall sign an undertaking to maintain the ram in good order and condition for two service seasons, and not to dispose of him during that period without the written consent of the County Galway Committe of Agriculture.

The annual show and sale:

Up to, and including, the 1957 season premium rams were selected and sold at only one approved sale, namely, the Annual Show and Sale of the Galway Sheep Breeders' Society held at Athenry early in September. In

1958 an auxiliary Show and Sale was held in Kilmaine[6] on the 2nd September, in addition to the main Annual Show and Sale of the Society which, this year, was held at Galway on the 4th September.
"At the Kilmaine Sale 74 registered Galway rams were selected for premiums and 53 of these were sold by public auction. At the Galway Show and Sale on September 4th, 108 rams were selected for premiums and 95 of these were sold by public auction. The total sales of premium rams realised £3,067. There was also a small sale of registered ewes held at Galway.
In addition to County Galway breeders, purchasers of premium rams came from neighbouring counties in which Sheep Breeding Schemes are in operation. The Department of Agriculture purchased a number of premium rams and had them located under a special scheme in the Aran Islands. Results from breeding with Galway registered rams located in the Islands in recent years has been very encouraging.

> As in previous years the Committee paid a subsidy of £75 to the Galway Sheep Breeders' Society to assist in their work for the improvement of Galway Sheep and for the publication of a Galway Sheep Flock Book."

The *Annual Report of the Galway County Committee of Agriculture* for 1962 included a fairly comprehensive article on The Sheep Enterprise. A section of the article - which was written appropriately in a parochial style - dealt with The Ewe Flock and portrays the trends in sheep breeding at that time:

> "What breed will you keep? At the moment there are approximately 350,000 ewes, mainly Galways, in this county alone. Even if it was considered desirable to replace these with some other breed or cross, it would be a huge undertaking which would take many years to accomplish. So whether we like it or not, for a considerable time to come, the breed of ewe in the county must be Galway. It is some consolation to know that the Galway has performed quite well against other breeds and crosses in recent trials. But from now on we must not be content

[6] Kilmaine is in County Mayo; a Show and Sale was held at Roundfort in 1967

> with any sort of Galway. We must have a ewe with class stamped on her. The year 1962 showed up the value of quality, both in ewes or in lambs and as time goes on the emphasis will be more and more towards quality.
> Avoid the big rangy sheep. Get a compact clean ewe with a good back and quarters, covered with a uniform coat of good wool. The ewe mated with a registered Galway ram, will produce grand ewe hoggets for breeding or, if mated with a Suffolk ram, she will produce fine cross-bred lambs suitable for the meat trade. You should be always selecting within your flock, culling the rough animals and the bad sucklers. You must get to know your good breeding ewes. Selling the best and keeping what's left is the worst possible type of breeding policy. Far too many of our good ewes are sold off at 3.5 years old. At that age they are just in their prime. Excessive teeth wear can be a problem and much work has been done on this aspect of sheep farming. There are many factors involved in teeth wear but it would appear that the most important factor in Galway is the very bare grazing in Spring. Wouldn't our trough feeding help to get over this a bit."

The article went on in a section on Store or Fat Lambs to advise:

> ...You have built up a business in stores over the years. Why not develop it further? With better feeding you will have forward lambs suitable for two marketsthe Midland and Northern feeders competing with the factory buyers. That will be to your advantage.

Thus, in 1962 farmers were being advised to "avoid the big rangy ewe". This brings to mind the distinction, already referred to, between Roscommons and Galways noted in 1932 by Joseph Hanly [xxvi]:

> "the principal breeds favoured are the Roscommon, a large white native sheep well suited to the midlands; and the Galway sheep, a white, compact animal of medium size, very hardy and prolific, recently established.

The 1962 advice conformed with a trend already in existence in 1958[xxvii] and probably in 1949, that is farmers in counties neighbouring on Galway were purchasing quality Galway rams for use on their white ewe flocks [xxviii].

Thus, in counties neighbouring on Galway a transition in commercial sheep flocks from Roscommon-type to Galway type was certainly happening in 1958 and possibly earlier.

There were other factors at work, that is a development of a market trade for early lamb and a demand by meat factories for well-finished lamb. The 1962 Galway Committee of Agriculture article continued:

"Should the lambs be "white" or Black-faced? ... The big problem is to maintain a balance in the county and there is room for both types of breeding. If we all go "black" where will we get our ewe replacements? Over a lengthy period, taking replacements and other things into consideration, there is not much in favour of one against the other. A lot will depend on the balance maintained between pure breeding and crossing. If ewe replacements are scarce the white breeding may pay better. It is not a good policy to see-saw between white and black ... If your grass and feeding are not good, then definitely stay all Galway. The disposal of the Galway wether lamb presents the big difficulty. He has a limited outlet as a store. The best outlet appears to be the carcase trade[xxix]. So get him into suitable condition as early as possible...

The Galway cross Suffolk is a popular factory lamb, and is also popular with Northern and Midland feeders who may want to finish them off at bigger weights for local markets. You have more opportunities with this cross[xxx]. It fleshes up quickly and develops into a mature type of carcase in 12 to 14 weeks, giving a 40 to 45 lb. carcase. The tail-end (that is, less well-developed lambs) can be finished off later in the year and the factory buyers will always be interested in those.More farmers are now taking their lambs to the factories and getting paid on a dead-weight basis but weights and prices are sometimes disappointing ... It takes a well finished lamb to kill out at 50%. A thin lamb will yield a carcase a long way short of that percentage ... The carcase must have a nice cover of flesh ... the big thin lamb may be down-graded, apart entirely from the weight it kills out at ... For crossing it is well to rely on compact, aged ewes with quality

... Put these ewes to a long, low Suffolk ram. Length is important. He will pass this factor on to his lambs. Try to get a ram with good quarters and broad flat kidney. A high prominent back-bone is hard to cover with meat and this fault in a ram may show up in the progeny ... The ram is half the flock ...".

The award of premiums:
In 1973, the Roscommon County Committee of Agriculture made provision for the award of premiums in respect of two sheep breeding schemes, i.e. Suffolk and Galway. The Annual Report of the Committee (1973) stated that in 1970 the GSBS introduced for the first time a system whereby the more prolific strains of the breed were identified under four classes, with different premiums.

Premiums were awarded and rams were purchased as follows:
(numbers for 1974 included also in third column)

Class	Premium £	Premiums Awarded 1973	1974
Class 1	20	2	4
Class 2	16	7	3
Class 3	12	8	8
Class 4	8	3	2

In 1980, Mr. James Greaney[xxxi], Chairman of the Galway Sheep Breeders' Society, sought to have "a special premium paid to (owners of) commercial Galway hoggets, bred from Pedigree Galway rams when those hoggets are retained for breeding".
In September of that year a *Workshop on Mixed Grazing* was held in the Great Southern Hotel, Galway. It was observed[xxxii] that "mixed grazing by cattle and sheep is widely practised on both low and hill grasslands in Ireland. The Galway-type sheep and cattle system is the dominant enterprise on the lowlands of Galway, South Mayo, South Roscommon, Tipperary (North Riding) and North Clare. The area involved is approximately two million acres. The traditional farming pattern in this

region involves the production of breeding ewes, and the sale of fat and store lambs, combined with the production of store cattle".

In 1988, Tom Sice, the Honorary Secretary of the Galway Sheep Breeders' Society, secured sponsorship from Patrick C. Joyce and Sons, Licensed Wool Exporters, of Church Road, Headford, Co. Galway, for publication of an illustrated promotional leaflet The Galway Breed: Ireland's Predominant Lowland Breeding Sheep. The leaflet advised that "The national flock of Galway type ewes was estimated at 269,000 in 1985 or 12% of the total flock of breeding ewes. The Galway- Suffolk cross produces excellent fat lamb and an excellent breeding ewe. For this reason it is important to at least maintain flock numbers at their present level. A Galway ram should be used to breed size and conformation back into your flock after a number of years of indiscriminate crossing. The Galway ram should also be used to maintain a nucleus of Galway type breeding ewes in your flock ... The Galway Sheep Breeders' Society was the first to start official recording[7] of pedigree sheep with the help of the Department of Agriculture and An Foras Talúntais ... Mating records, litter, lamb growth rate and mortality are recorded for all pedigree ewes. The litter size records are used to calculate an index for ewes & their progeny ... Premium Class I rams are from ewes with a consistent history of twinning. Class II rams are from ewes with an average litter size of at least 1.5 ... In 1986, 85% of rams entered for sale were Grade I ...
The prolificacy of the breed is exemplified by the performance now attained by the pedigree flocks. Average litter size for 1986 and 1987 was 1.71 and 1.68 respectively ...
The following table gives the litter size attained by the four leading flocks each year.

Flock Placing	*1984*	*1985*	*1986*
1st.	1.97	2.0	1.9
2nd	1.77	1.86	1.89
3rd	1.71	1.79	1.84
4th	1.71	1.78	1.83

[7] in 1969

The following table indicates the ewe performance required for each premium class.

Number of lambing records per ewe	Total lambs needed for *Class I*	*Class II*
1	3	2
2	4	3
3	6	5
4	7	6

Only sons of Class 1 ewes are used as stock rams by pedigree breeders and only daughters of Class 1 and Class 2 ewes are now eligible for registration in the Flock Book.

The average litter size of all recorded ewes in 1986 and 1987 was as follows:

	1986	1987
2-year old ewes	1.66	1.57
3-year old ewes	1.73	1.74
4-year old ewes	1.74	1.73
Mean values	1.71	1.68

The leaflet also advised that the growth rate of lambs is as good as any other breed or cross with lambs reaching 37 kg by 100 days of age. The pre-weaning growth rate of all lambs recorded in 1975 to 1978 inclusive was 283 g/day. An average growth rate of 311 g/day was achieved by the top third flocks over the same period.

The ewe bodyweight was given as:

18 month old ewes	63 kg
30 month old ewes	69 kg
42 month old ewes	74 kg

Rare Breed Status:

The writer of this present volume found that all of the performance figures set out in the leaflet are a very realistic guide and were in the range achieved by his small flock at Tara in County Meath in the 1990s. In 1994, the Irish Genetic Resources Conservation Trust Ltd.[xxxiii] (IGRCT) issued two leaflets[xxxiv] drawing attention to the declining numbers of pure-bred

Galway sheep. In its bi-lingual (Irish/English) leaflet it listed the classification system[8] used by the UK-based Rare Breeds' Survival Trust for sheep breeds needing conservation effort:

At Risk	1,500 survivors;	Endangered	500 survivors
Vulnerable	900 survivors;	Critical	300 survivors.

The IGRCT leaflet went on to state "In March, 1994, the Pedigree Galway sheep population had reached critical status with 300 remaining pedigree breeding ewes. This number of breeding ewes, with the small number of working rams it implies, represents an extremely narrow genetic base for the future existence of this native breed".

Unknown to the writer there were, in fact, many more pure-bred Galway sheep in existence at that time, but their owners had ceased to take part in the activities of the Galway Sheep Breeders' Society. It may have taken but a short time for their Galway flocks to have been dispersed or inter-bred with sheep of other breeds.

It was traditional for meetings of the breed Society to be held regularly at Mr. Thompson's premises in Athenry. During the mid-1990s, concern was evident at those meetings about a declining interest amongst farmers in the breed. The Society set about attempts to reverse the trend. Some publicity was given to the decline in numbers of Galway sheep and this soon led to the breed being officially classified as "endangered" by the Department of Agriculture (an inventory was published in 1998).

A letter, dated 7th March 1994, to the writer from Dr. Jim Flanagan of the Department of Agriculture, Food and Forestry, stated: "As regards rare breed status of Galway sheep we are endeavouring to get them included in the REPS measure and to this end have responded to a series of questions from the EU Commission in recent weeks. The matter is likely to come before the STAR Committee in Brussels this month for consideration.

[8] A global view of conservation requires a different focus. The terms "endangered", "critical", etc. are defined within narrower limits in FAO Publications: The Global Strategy for the Management of Farm Animal Genetic Resources: Executive Brief and A Call for Action, 1992.

As a gesture of encouragement to Purebred Galway breeders to participate in the flock-recording scheme we have waived the recording fee for this year. Eight of the ten recording last year have rejoined and we have eight new flockowners showing some interest". This official action provided owners of Galway Sheep with an opportunity to secure a supplementary income through premiums payable for conservation of endangered breeds under the E.U's Rural Environmental Protection Scheme (REPS). "Financial assistance equivalent to 100 ECUs per livestock unit[9] is paid for in-situ conservation." The conservation premium is additional to the Ewe Premium which is payable in all parts of the country and is intended to supplement producers' incomes by compensating them for low market prices. Premiums are 100% financed by the EU. The value of the Ewe Premium for a given year depends on market prices during that year and is not finally fixed until March of the following year. Figures[xxxv] for the year 1991, when the value of the premium was £22.69 per head, may serve as an illustration. The payment was made in 3 instalments:

First Advance;	30% in July 1991 - £6.75
Second Advance;	30% in Nov. 1991 - £6.89
Balance;	40% in April 1992- £9.05

More specifically, the development of the *Galway Sheep Conservation Project* marked a milestone in efforts to secure the future of the breed. "It entails the development of a computerised flock book, improved identification and registration of breeding stock, identification of further flock owners interested in maintaining registered sheep, drawing up mating plans which will conserve genetic variability and identifying suitable rams for cryopreservation (of semen)." The project is led by Dr. J. P. . Hanrahan, Teagasc, Athenry, Co. Galway.

In 1994, the membership of the Galway Sheep Breeders' Society had declined to ten. However, the work undertaken by those ten - especially by Mr. Tom Sice, the Honorary Secretary - resulted in a dramatic resurgence in membership and an increase in sheep-numbers in the pure-bred register, from 271 ewes (14 flocks) in 1995 to 734* ewes (38 flocks)

[9] One sheep = 0.15 livestock units.

in 1998. [*263 were said to be of unknown parentage and newly recruited.] Forty-three members of the Society were listed in the catalogue produced for the Show and Sale at Athenry in September 1999. As noted above, Dr. J. P. Hanrahan of Teagasc (Athenry) became very much involved with the Society from 1995 and assisted the Chairman (Mr. Joe Casey) and officers to revise the rules governing the organisation and adopt a revised sheep-breeding policy. As a result breeders are provided with a list of their lambs born in the previous season set out in a merit index[xxxvi] to form a basis for selection of future dams and sires. A small example is given here for some ewes in Flock LC from the Report for 1999 [similar data are available for lambs]:

Tag No.	Year of Birth	Litter Size	Premium Class	Merit Index	Mated to Ram
LCL1	'95	2	4	-7.817	JMH3
LCL3	'95	3	1	10.363	JMH3
LCL7	'95	2	3	-2.500	JMH3
LCN4	'97	2	2	4.000	JMH3

The Society needs to take effective steps to harness this kind of information to raise fecundity levels, but it also needs to take action to retain superior stock that carry the traditional, outstanding, conformation and wool qualities of the breed. In fact, there is something of a dilemma facing breeders who attempt to enhance the commercial qualities of their stock through intense selection: the genetic base of flocks is narrowed - that is, there is a loss of genes - whereas conservationists argue that there is a need to retain a broad genetic base and ensure the established identity of each individual breed. The theoretical conflict may be resolved by conserving the more traditional strains in heritage parks. Such an approach would require a planned long-term strategy. That strategy should also embrace the idea of a wide territorial dispersal of flocks to reduce the risks of extinction that arise from possible local outbreaks of highly infectious disease, like foot-and-mouth disease.

Owners with winning rams at GSBS Show and Sale 1996. From left: Tom Sice (with PSL 6), Tom Murphy (MML8), Martin Fallon (BFL4), Paddy Sice (PSL3) and Bernard Fallon (with champion BFL6 bought by Joe Casey)

Henry Hughes (wool expert, right) with from left: Gerry Nolan (Ardrahan), Michael Conneely (Monivea), Joe Casey (Ballyvaughan)

Children Michael & Jacinta Higgins (Athenry) with from left: Paddy Sice, Tom Sice, Michael Maher, Seamus Cuddy (Ulster Bank) Joe Casey, Jack Casey and Martin Fallon, 1997

Young pedigree Galway ewe attends to her new born lamb at Tara, 2000

— CHAPTER FIVE —

GALWAYS BACK FROM THE BRINK

A Note in the *Irish Farmers' Journal*

Under the above chapter heading the Irish Farmers' Journal of October 19, 1996 (page 16) reported on the resurgence of interest in the breed. The article[10] stated:

"On September 16 last the Galway Sheep Breed Society held its 73rd annual show and sale at Athenry Mart. A lot was at stake for Ireland's most famous native breed. Two years ago the breed was facing extinction; last year's sale was a flop.

Much to the relief of the breed's supporters, (this year) there was a full clearance of the almost 100 sheep on offer.

- 35 hogget rams made from 150 to 505 gns; Joe Casey Murrough, Clare, Chairman of the Society, bought the top priced ram for 505 gns.
- 20 ram lambs made from 100 to 300 gns.
- 30 registered ewes sold from 120 to 155 gns.

Two years ago [steps were taken to encourage a revival of interest in the breed] ...Meetings were held and lapsed breeders were encouraged to register their stock. The flock book was brought up to date.

Teagasc sheep specialist, Gerry Scully, and breed experts were assigned to examine sheep for inclusion in the flock book.

Laurence Alderson, the chairman of the Rare Breeds Survival Trust had persuaded Mr. Alan Cheese of Newcastle-on-Tyne to buy 18 in-lamb Galway ewes and a ram in 1990. There are also a number of small flocks of Galway ewes in Northern Ireland. The Galway breed (Society) was established in 1922. It was preceded by the Roscommon breed, which was established in 1895. One, Nicholas Cotter, took an enormous interest in the Roscommon breed, and the breed was allowed to die out with him. One of the big promoters of the Galway breed was Oranmore breeder Christy Kearns (Kerin). He was chairman of the breed (Society) for an extended period. It is reckoned that some bloodlines from the Lincoln

[10] The author has made some minor corrections to the content of the article.

Longwool and Ryeland were incorporated at the start-up of the Galway breed and that this accounts for the high quality wool. Over the years there was a lot of emphasis on long wool and big sheep, to the expense of prolificacy in the Galway breed. However, the breeders are reversing the trend towards single lambs, and have breeding programmes to increase prolificacy.
The cross between the Galway and the Suffolk has proved very popular in the West. This cross can only be perpetuated if there are Galways available for breeding."

This mention of cross-breeding in the *Journal* article reminds us that in its 1988 leaflet the Galway Sheep Breeders' Society outlined the practices of crossbreeding:

"The Galway Ram is an excellent choice for use on hill breeds for the production of Crossbred Ewes. The Galway/Blackface Mountain Ewe in particular is ideal for this purpose and commands a premium price in the market place.
The Galway Ewe makes an excellent fat lamb mother and is widely used in crossing with Down Type and Texel Rams for the production of early and mid-season lamb".

This reference to Texels marks a relatively new trend in the development of Galway crossbreds to meet market requirements. It must be emphasised that the production of crossbreds for the food chain, including the demanding French market, is one thing: the maintenance of pedigree Galways as a distinct breed, free from Texel "blood" is another. The breeders must be vigilant and exercise rigorous culling to avoid any ingress from other white breeds, such as the Texel[xxxvii].

The European Influence

In 1984, a working party of the European Agricultural Animal Production Commission produced its final report on the Conservation of Animal Genetic Resources in Europe[11]. In the cattle section, Dexter, Kerry and Irish Moyled were entered for consideration under a Great Britain

[11] Maijala, K. *et al*, 1984, *Livestock Production Science, 11; 3-32*

reference point. In the sheep section there was no entry for an Irish breed although a total of 63 European breeds were listed. The report was presented under two main criteria (a) prevention of loss of breeds, and (b) prevention of loss of genes within breeds.

In 1995, a meeting was held in Agriculture House, Dublin, to finalise the Country Report on genetic resources for food and agriculture for submission to the IPGRI by the end of March. About that time also a new Irish Pedigree Sheep Breed Improvement Programme commenced. Its objectives were to improve growth rate and carcass quality of pedigree sheep breeds and record growth rates of all lambs in participating flocks. The growth rates and ultrasonic measurements of fat were to be combined to provide a "Lean Meat Index" for all ram lambs. The ultimate objective was to encourage pedigree and commercial breeders to use the performance records to select ewes and stock rams for further breeding. The documentation invited applications from members of a recognised Sheep Breed Society, but did not specifically mention Galways.

While the Department of Agriculture Food and Forestry (DAFF) has direct responsibility for the conservation and utilisation of genetic resources for food and agriculture it is aided and advised by a Committee[xxxviii] on Genetic Resources for Food and Agriculture which was established in 1996.

The role of Non-Governmental Organisations

The government has a duty to provide opportunities and incentives to ensure the long-term conservation of our native breeds of farm livestock. Factors of history determined that only a small number of breeds were officially recognised from within the indigenous livestock population on this island. This contrasts sharply with Britain where a great range of breeds (at least 33 sheep breeds) emerged and became established. Here, some local populations of sheep, cattle and equines became recognisably distinct but few had a framework of support from organised groups or societies to justify their official designation as breeds. Where such societies exist they need adequate technical support and constant encouragement to safeguard the long-term survival of the livestock that are their special concern.The problems that beset the pedigree Galway sheep population during the 1980s and 1990s illustrate some of the dilemmas facing conservationists. The flock-owners make their living in the world of commercial farming and that world is constantly changing because of demands in the market-place. A response to those demands can put considerable selection pressures on flocks and on the segment of biodiversity that they represent. The GSBS is well established and well managed and is quite capable of looking after its own affairs.Like many other independent non-government bodies, it needs state support, as well as advice and encouragement from relevant outside sources (e.g., IGRCT & the IRBS[12]) to safeguard its commodity and its defined objectives. All the signs now indicate that a policy of sustained support is in place, and Ireland's rare breeds are set to continue as part of the rural landscape in the third millenium.

The Galway Sheep Breeders' Society of Great Britain
[Internet Web site: www.sheep.ukf.net]
The following information was supplied on 30th April, 2001, by the Secretary of this relatively new body, Mr. Alan Cheese, Hill Croft, Ingoe, Matfen, Newcastle-upon-Tyne, NE 20 OSJ.
Telephone: 01661 886408
Fax: 087 070 60209.
Chairman: Mr John Heathcote Ball, Ulverscroft Lodge, Ulverscroft, Markfield, Leicester.
The first pedigree Galways were imported to the United Kingdom by Alan and Dorthe Cheese in 1990. A selection of 29 high [merit] index sheep, three males and twenty six females, was made from the flocks of Tom Sice, the late Bernard Fallon, Joe Casey, the late Jimmy Greaney, Martin Fitzpatrick, Paddy Broderick and the Maher Brothers. Gerry Scully (Teagasc) very kindly purchased one of the rams for the import at the annual breed sale. Such was the foundation of the "Mavesyn" flock, which, in its first year, secured a prize in the flock competition in the breed's home country!

The "Mavesyn" flock continued until 1994, when it was taken on, in its entirety, by John Heathcote Ball, under the "Ulverscroft" prefix. In 1997, Mr Heathcote Ball also acquired the major part of the "Dal" flock from Mrs Annabel Holt. Mrs Holt's flock had been founded with some of the

[12] Secretary, Michael de las Casas, Larch Hill, Kilcock, Co. Kildare.

first lambs from the "Mavesyn" flock.
A further 27 ewes and three rams were imported in 1998 by Roy Tucker and Kelvin Yue, with stock coming from Healey, Sice, Maher and Connolly flocks. Pedigrees of all pure-bred Galways since the original import have been recorded, the flock book being maintained by Alan Cheese, with help from John Ball and Annabel Holt. Regular sales from "Ulverscroft" and "Dal" resulted in the establishment of a number of new flocks and by 2000 there were sufficient flocks to make the formation of a breed society feasible. This took place early in 2000 and there are now 17 flocks spread from Aberdeen to Devon and North Wales to Lincolnshire. The British population of registered Galways is approximately 130 ewes.
There is considerable interest in both Galway lamb and mutton from specialist high [grade] meat retailers. Crossing of the Galway with native breeds is also beginning.
The Galway is listed as Category 2 (Endangered) on the Rare Breeds Survival Trust's priority list and the Society has made the RBST Show and Sale, held at Stoneleigh, Warwickshire, in September, its official breed sale. Each year the Champion Galway at Stoneleigh is awarded the Jimmy Greaney Memorial Trophy, in recognition of the help Jimmy gave in establishing the breed in the U.K.
[Note: The author of this volume is pleased to observe that the 1998 importations included two females (MMP17 and MMP18) bred by the author at Macetown, Tara, Co. Meath, sold as lambs to Tom Murphy of Shralea, Creagh, Ballinasloe, and then sold to Kelvin Yue in England. The Galway G.B. News is a newsletter published by the Society in Great Britain]

Joe Casey (left) Chairman of the Galway Sheep Breeders' Society, receives trophy for his prize-winning ram from the staff of Chanelle Veterinary with judge Pat Loughnane (right)

This Galway ram was Longwool Breed Champion at the Rare Breeds Survival Trust Show at Stoneleigh (Warwickshire) in 2000

Traditional póirín permitting sheep to move from field to field but restricting cattle (Source P.J. Daly 1966)

Traditional Sheep-cock, with commercial Galway Sheep (Source P.J. Daly 1966)

APPENDICES:

More Than A Century of Voluntary Support

••••

APPENDIX - I

The Roscommon Sheep Breeders' Association -1896
HONORARY OFFICERS

President

THE RIGHT HON. THE O'CONNOR DON, P.C.

Vice-Presidents.

Right Hon. Arthur Baron De Freyne, D.L.
Major M.J. Balfe, D.L.
Count Arthur Moore, D.L.
Luke J. Irwin, Esq.
Arthur Hugh Smith-Barry, Right Hon., M.P.
John L. Cotton, Esq.
Captain Thomas Longworth Dames, J.P.
Michael Kelly, Esq., J.P.
John Keane, Esq.
Thomas Rorke, Esq., J.P.
Thomas Conry, Esq.
Andrew Ganly, Esq., V.S.
John J. Ryan, Esq.
Richard Hague, Esq., J.P.
Terence Glancy, Esq.
T. A. P. Mapother, Esq., J.P.
Hector, J.C. Toler-Aylward, Esq., D.L.
Francis A. Flanagan, Esq.
Richard Ganly, Esq.
M. J. Cleary, Esq., V.S.

Committee of Management

Major M. J. Balfe, D.L, J.P.
Timothy Rorke, Esq., J.P.
Francis A. Flanagan, Esq.,
Richard Hague, Esq.
John L. Cotton, Esq.

Editing Committee

Major M. J. Balfe, D.L. (Chairman)
John L. Cotton, Esq.,
John Ryan, Esq.
Timothy Rorke Esq., J.P.
James Ganly, Esq.
Joseph Banahan, Esq.
Luke J. Irwin, Esq., J.P.
John O'Connor, Esq.

Inspection Committee

Timothy Rorke, Esq., J.P. (Chairman)
John L. Cotton, Esq.,
Patrick P. Flanagan, Esq.
John Keane, Esq.,
James Ganly, Esq.
Joseph Banahan, Esq.,
John O'Connor, Esq.,
Frederick P. Smith, Esq., J.P.

Finance Committee

Major M. J. Balfe, D.L. (Chairman)
Richard Hague, Esq., J.P.,
Luke J. Irwin, Esq., J.P.
Timothy Rorke, Esq., J.P.,
Francis A. Flanagan, Esq.
Michael Kelly, Esq., J.P.,
Cox Cotton, Esq.

Trustees

Michael Kelly, Esq., J.P.
Francis A. Flanagan, Esq.
Terence Glancy, Esq.,
Michael Mitchell, Esq.

Honorary Secretary and Treasurer

Matthew Flanagan, Esq., J.P.,
Tomona, Tulsk, Co. Roscommon.

Members of Council

John Ganly, Esq., (Chairman)
Joseph Banahan, Esq.
Matthew Flanagan, Esq., J.P.
Captain T. Longworth-Dames, D.L.
William Neary, Esq, J.P.
Thomas J. White, Esq.
Michael Mitchell, Esq.
M. G. Sweeney, Esq., J.P.
Timothy Rorke, Esq., J.P.
Cox Cotton, Esq.
Richard Flynn, Esq.,
Patrick P. Flanagan, Esq.,
Michael Rorke, Esq.
Wm. Holmes, Esq.,
Mathew Gately, Esq.

APPENDIX - II
The Galway Sheep Breeders' Association

The members of the Galway Sheep Breeders' Society agreed in 1999 to adopt a new structure for the old organisation in order to meet the requirements of REPS. From March 2000 it became a legal entity as a registered company called the Galway Sheep Breeders' Association Ltd. One noteworthy change was an escalation in the costs of active membership of the organisation. This change came as a surprise to the author of this volume. All members, including those interested in breeding pedigree Galways as a conservation - rather than as a commercial - endeavour, now have to pay a variety of fees in addition to a membership fee of £30. Fees for entry in the Flock Book are £3 for females and £6 for males. The fee for entry of females in the Supplementary Register (Annexe) is £5. An annual fee of £2 per ewe will be collected for ewes entered on the Mating List. The fee for issuing a certificate will be £2. These fees will be revised regularly by the Council.

At its meeting on 22 February 2001 the members expressed much concern about "the huge decline in the number of registered members from (the year) 2000 to 2001" and "the narrowing of blood lines once again after the major effort made by members of the society in up-grading sheep in the mid 90's". Another stated concern is that "the Galway sheep breeders' society set up and became a limited company in '99 - Galway Sheep Breeders Association Ltd. There was a big cost involved. Was the changeover necessary?" The author sympathises with those expressing such concerns. It is noteworthy that the word "conservation" was inserted as an afterthought in the rules governing the new Association. Many of those interested in livestock conservation are not necessarily motivated by the prospects of profit, but they are dissuaded by the imposition of unnecessary costs. The number of pedigree Galway ewes mated in 1999/2000 was 850; the number mated in 2000/2001 was 550 - a decline of 35% in one year; A major part of rare breed conservation requires that there be as many flocks as possible so that inbreeding is reduced and diversity increased. Thus, many owners need to be enticed into continuing membership.

Officers of the Galway Sheep Breeders' Association in 2001

Chairman:
Mr. Joe Casey, Murrough, Ballyvaughan, Co, Clare.

Secretary:
Mr. Thomas Sice, Boyhill, Athenry, Co. Galway.

Assistant Secretary:
Mr. Tom Murphy, Shralea, Creagh, Ballinasloe, Co. Galway

Directors:
Mr. Fintan Connolly, Rathbaun Farm, Ardrahan, Co. Galway
Mr. Gerry Nolan, Coxtown Ardrahan, Co. Galway
Mr. Thomas Maher, Belville, Monivea, Athenry, Co. Galway
Mr. Michael Conneely, Tulicar, Monivea, Co. Galway

Ordinary Members:
Twenty are listed in addition to the seven officers. One member lives in County Cork, two in County Tipperary, one in County Louth, one in County Roscommon, two in Northern Ireland, two in the U.K. (Devon and Somerset), and eleven in County Galway.

[Note added in print: The meeting for 2001 of the Advisory Committee on Genetic Resources for Food and Agriculture was held on 29th May in Agriculture House. The author drew the attention of the Committee to the recent decline in membership of the Galway Sheep Breeders' Association (formerly, 'Society'). He suggested that the reasons for the decline were deserving of investigation, and requested that special attention be given, as a matter of urgency, to the conservation of diversity within the Galway breed. The Chairman, Dr. Teehan, agreed that there was a need to take appropriate action.]

END-NOTES (including bibliography)

i P. J. Daly, 1966 Sheep Husbandry Survey in Kilmaine, AFT

ii FAO 1992, A Global Strategy for the Management of Farm Animal Genetic Resources: Executive Brief and Call for Action, 1992

iii Sketches of the Merino Factory: Descriptive of its Origin and Progress and of its system of Discipline and Moral Government (presented to the Dublin Society, 19th December, 1816) Dublin 1818.

iv Corrigan, Andrew, 1859, Theory and Practice of Agric. 3rd ed. Dublin.

v Youatt, W. 1900 The Complete Grazier and Farmers' and Cattle-breeders' Assistant, 14th edition by Wm. Fream,Crosby Lockwood and Son, London

vi Gloucestershire Ryelanders, see Young's Annals (1793) vol. 20: p. 17

vii Roscommon Flock Book, vol 1, 1896

viii Murphy, E., 1849, The Agricultural Instructor, McGlashan, Dublin.

ix Baldwin, Thomas, 1881, Agricultural Class-Book, 3rd ed. Alex. Thom and Co., Dublin

x The Standard Cyclopedia of Modern Agriculture and Rural Economy (12 volumes), 1910, The Gresham Publishing Company

xi Arthur Young 1793, Annals 20 p. 37 refers to Lincoln/Spanish crosses at Spilsby, Lincolnshire.

xii Saorstat Eireann: AgricultureA note on some outstanding features of the Irish Free State's Agricultural Resources:issued by the Department of Agriculture of Saorstat Eireann

xiii Saorstat Eireann Irish Free StateOfficial Handbook, The Talbot Press, Dublin, 1932

xiv Anon, 1958, Sheep Raising, pp. 56-60 in Annual Report of the Roscommon County Committee of Agriculture.

xv Cotter, N. P., 1939, Roscommon Sheep, Journal of the Department of Agriculture (Ireland)

xvi Pat Boland, 1960, Sheep, Ceres (Educational Supplement to Students' Magazine), Faculty of Agric., UCD, pages 5 - 19.

xvii Source: Leicester Longwool Sheep Breeders' Association, Leaflet, 1994

xviii Hunter, A., 1803, Georgical Essays, Wilson and Spence, York.

xix 24th General Report of the Department of Agriculture and Technical

Instruction for Ireland (1923) page 67

xx Christy Kerin was one of 67 witnesses who gave evidence to the Inter-Departmental Committee on the Cleanliness and Wholesomeness of the Milk Supply - Report 1928. He was an agent (supplier/distributor) for Guinness; a nominator for the annual Clounanna Coursing event, and a member of the Galway County Committee of Agriculture.

xxi Report in Connacht Tribune (page 5) of 31/03/1923

xxii Report in Connacht Tribune (page 5) of 16/06/1923.

xxiii Annual Reports of the Minister for Agriculture, 1930s.

xxiv AGM Report in Farmers' Gazette, XCVI, No. 11, p.254, March12, '38

xxv Developments in Sheep Husbandry, page 34 in Annual Report of Galway County Committee of Agriculture, 1964.

xxvi Saorstat Eireann Irish Free State Official Handbook, The Talbot Press, Dublin 1932

xxvii Annual Report of the Galway County Committee of Agriculture for 1958

xxviii "Galway, Mayo, Roscommon and Clare County Committee of Agriculture give premiums to commercial sheep producers who purchase selected registered rams. Galway County Committee give 100 premiums, value £5 each." Report for 1967 of Galway County Committee of Agriculture

xxix as opposed to the store trade

xxx than with Oxford Down crosses

xxxi See Annual Report of Galway County Committee of Agriculture, 1980

xxxii Young, F., Mulligan, S., Hunt, S. and J. Whiriskey (ACOT Roscommon), 1980, Mixed Grazing Extension Work in Co. Roscommon.

xxxiii The Irish Genetic Resources Conservation Trust Ltd. was founded on 22nd March, 1994 and now operates from an address in Dublin: c/o Trinity College Botanic Gardens, Palmerston Park, Dartry, Dublin 6. The Hon. Sec. is Dr. Deirdre Lynn, School of Botany, Trinity College, Dublin 2: Email: delynn@tcd.ie

xxxiv Galway Sheep and Caora na Gaillimhe

xxxv Taken from The IFA Guide to grants, premiums and headage payments, April 1992.

xxxvi The method of calculation of the merit index is set out as rule 16 of the Rules of the Galway Sheep Breeders' Association Ltd. 2000/2001

xxxvii Texel (source - Sheep Breeder and Sheepman Magazine., Dec 1993, page 76) Originally from Netherlands, introduced into US in 1990. The Texel will produce a leaner lamb (than Rambouillet types) but the lamb must be slaughtered at a lighter weight to get the full effect.(A Texel-sired lamb marketed at 125-135lb will be too fat, according to R. M. Jordan, page 30, a 100 lb lamb leaner, more tender and tastier.)

xxxviii Members of the Committee on Genetic Resources for Food, Agriculture and Forestry are: Dr. Tom Teehan, DAFF, Chairman; Dr. Leo Curran, Tara, Co. Meath; Mr. John Dillon, IFA, Irish Farm Centre, Dublin 12; Dr. Seamus Hanrahan, Athenry Research Station, Teagasc; Dr. Rory Harrington, National Parks and Wildlife Service; Ms. Anita Hayes, Irish Seed Savers Assoc., Capparoe, Scariff, Co. Clare; Prof. Michael Hennerty, UCD; Dr. Diarmuid McAree, Forest Service, Dept. of Marine and Natural Resources; Dr. Maurice O'Riordan, ICMSA, Limerick; Dr. Steve Waldren, TCD Botanic Gardens; Prof. Edward Walsh, UCD; Mr. Donal Coleman, DAFF, Backweston (Technical Adviser); Mr. Liam Feeney, DAFF (Secretary; - the earlier Secretary was Mr. Karl Kelly).